MILLION DOLLAR SECRETS
of a *Successful* Inventor/Attorney

by Clemens V. Hedeen, Jr.

Copyright © 2017 Clemens V. Hedeen, Jr.

All rights reserved. No part of this book shall be reproduced, stored in a retrieval system, or transmitted by any means, electronic, mechanical, photocopying, recording, or otherwise, without written permission from the publisher. No patent liability is assumed with respect to the use of the information contained herein. Although every precaution has been taken in the preparation of this book, the publisher and author assume no responsibility for errors or omissions. Neither is any liability assumed for damages resulting from the use of information contained herein.

Note: This publication contains the opinions and ideas of its author. It is intended to provide helpful and informative material on the subject matter covered. It is sold with the understanding that the author is not engaged in rendering professional services in the book. If the reader requires personal assistance or advice, a competent professional should be consulted.

The author specifically disclaims any responsibility for any liability, loss, or risk, personal or otherwise, which is incurred as a consequence, directly or indirectly, of the use and application of any of the contents of this book.

Clemens V. Hedeen, Jr.
Attorney at Law
DBA Hedeen International, LLC
228 N 14th Ave
Sturgeon Bay, WI 54235
920-743-7225

ISBN: 1978132670
ISBN-13: 978-1978132672

MILLION DOLLAR SECRETS OF A SUCCESSFUL INVENTOR/ATTORNEY

Secrets of a Successful Inventor/Attorney who conceived two billion-dollar toy concepts…
Micro Machines™ and the Nerf Dart Gun™.

This book is a must-read in order to save inventors thousands and thousands in time and money on their quest to protect their ideas and have their concepts made, marketed, and sold by companies or by themselves.

Contents

Introduction 9

Acknowledgements 11

Preface 13

Chapter One - The Wacky World of Inventing 15

Chapter Two - Finding Your Way Through Uncharted Territory 17

Chapter Three - You've Got the Best Idea Since the Invention of the "Rubber Grape!" 21

Chapter Four - To Prototype or Not to Protoype – That is the Question! 27

Chapter Five - How do I "Kevlar" My Idea? 31

Chapter Six - Do I Need Professional Help? 33

Chapter Seven - Hire an Attorney, or Swim With the Sharks 37

Chapter Eight - Dealing With Big Corporations – the 600 Pound Gorillas 41

Chapter Nine - Holy Cow: They Like My Idea! 45

Chapter Ten – The Product Licensing Agreement: Negotiating With "The Devil" 49

Chapter Eleven - Tracking Your Product... Stay Awake and Keep Your Eyes Open..55

Chapter Twelve – The 9 Biggest Mistakes You Can Make as an Inventor 59

Chapter Thirteen – Are You Crazy Enough to do it Yourself? 63

Chapter Fourteen - The 9 Best Things I've Done as an Inventor 67

Chapter Fifteen - Summary..71

Special Preview of Am I Crazy? No. I'm a Toy Inventor

....73

Exhibit A – Confidential Disclosure Agreement (CDA) 85

Exhibit B – Confidential Disclosure Agreement (CDA) 89

Exhibit C – Non-Confidential Disclosure Agreement (NCDA) 93

Exhibit D – Option Agreement...95

Exhibit E – License Agreement..97

Introduction

Million Dollar Secrets of a Successful Inventor /Attorney is written by Clemens V. Hedeen, Jr., an inventor who conceived two billion-dollar toy concepts... Micro Machines™ and the Nerf Dart Gun™.

Hedeen is an attorney and inventor, who has been inventing toys for the past 37 years. He has dealt with the mega toy companies such as Hasbro and Mattel. He and his wife Kay Lee operate a toy inventing business called Fun City USA. They have even done self-manufacturing.

Over the years Hedeen has produced two billion-dollar mega toy hits. The first was Micro Machines which became the #1 toy in the world and which sold over a billion micro sized toy cars and accessories.

The second was the Nerf gun and soft darts. Hedeen sold this concept to Kenner Toys which was later purchased by Hasbro. Hedeen's original gun mechanism and soft dart concepts are still being sold by Hasbro. Total sales of this line are in the billions.

Hedeen offers a rare view into the professional world of inventing. This book discloses the secrets of the inventing business from A to Z.

This book starts with helping would-be inventors throughout the creative process. It then moves on to giving insight on what to do with your invention.

From prototyping, to finding a manufacturer, to self-manufacturing, to sales and marketing, this book goes over it all in detail.

Some of Hedeen's best secrets come from his experience as an attorney in dealing with patents, copyrights, and trademarks and extends to licensing contracts and related issues. Some of the other issues include auditing, sales, liability, insurance and arbitration or litigation.

Everyone who has a great idea for an invention should read this book. With forty plus years of legal and inventing experience, Hedeen offers invaluable advice to amateur and would-be professional inventors. He discloses numerous secrets of the trade!

ACKNOWLEDGEMENTS

I want to acknowledge my wife, Kay Lee, who is also my partner in our inventing business. She is the glue that keeps it all together.

Not only is she creative but she brings her intuition and practical nature to our inventing and development company.

She has encouraged me to write about our experiences and share our secrets and the "tricks of the trade." She agreed to be my editor and without her encouragement and expertise this book would not have been written.

I also want to acknowledge my three children Justus, Nikki and Barret who have all suggested I help others by writing this book. Also, my step-daughters Tiffany and Antonia are always there to encourage me.

My true passion wasn't discovered until I fell into inventing. Looking for a new venture, I had the opportunity to start a toy inventing company.

Inventing opened up a whole new world for me. It was at the same time exciting and challenging, exhilarating and frustrating, financially rewarding and yet very expensive.

My goal is to share my experience in the wacky world of inventing with you. Hopefully my knowledge will save you some time and money as you venture into the mysterious land of creating and inventing.

PREFACE

I started my professional career as a trial attorney in Milwaukee and Green Bay, Wisconsin. I received my law degree from the University of Wisconsin. Trial law seemed a natural since one of my hobbies was acting with the Wisconsin Players. Later my legal career included the positions of District Attorney, Corporate Counsel, and Family Court Commissioner.

However, my real love was creating something from nothing. I started with real estate development, building apartments, restaurants, and even a bowling center. Later I moved on to condominium and resort development. Eventually I found my true passion which was creating toys for children of all ages.

CHAPTER ONE

THE WACKY WORLD OF INVENTING

Anyone can be an inventor. Millions of people already are inventors. For the most part, these are people who have an idea to solve a problem, a personal problem, that's been annoying them. Then what do they do? Nothing. Because they don't have the knowledge, money, or ambition to take the next steps. They are reluctant to enter the "wacky world of inventing."

Thus, most of these inventors have great ideas that languish and die. Or someone else beats them to the market. How often have you heard people say when confronted with a new product, "Gee I had that idea years ago, but I didn't do anything with it"!

Of course, not every invention idea is worthy. In fact, the great majority are probably worthless. How do you know which one is "gold" and which one is "quick-silver"? Fortunately, I have been able to create and invent many financially successful toy concepts with the help of many other talented and creative people, including my wife Kay Lee.

The Micro Machines concept came to me during an idea session in our toy shop. The idea was to micro size detailed vehicles and to create a whole new micro world. Instead of selling one car at a time we would sell five or six in a package and kids could carry them around in their pockets. I sold this concept to Galoob Toys and it caught on like wildfire. Micro Machines had a great run and was the number one selling toy in the world for a while and was in the top ten for years. Galoob eventually sold almost a billion Micro Machines. It was also one of the first really successful micro toy lines. The success of Micro Machines spawned many other successful micro toy lines.

However, for every successful toy that we've created there are <u>twenty or thirty</u> that never make it. The road to success through the "wacky and wild world" of inventing takes many twists and turns. Every financially (commercially) successful invention has an element of luck attached to it. With knowledge, you can increase your chances of success and minimize the part that luck plays in your path to fame and fortune.

The purpose of this book is to help inventors navigate through the treacherous waters of inventing. There are pirates who would love to steal your idea. There are dangerous winds that could blow you in the wrong direction. There are "eddies" that could lull you into inaction. There are merchants who would have you purchase useless goods for your long voyage, while they profit and you flounder.

CHAPTER TWO

FINDING YOUR WAY THROUGH UNCHARTED TERRITORY

The first thing you need to do is to figure out what you have and where you are at. One definition of an invention is "something thought up or mentally fabricated." An invention is created by "ingenuity or creativity." (Webster's New World Dictionary)

Your invention may simply be in the idea form or you may have already progressed to a working prototype. Millions of people never get past this first stage. They never get past the conception stage.

Why do so many people have creative ideas? Because it's the nature of human beings to solve problems by creative solutions. This is the history of man from the invention of the wheel to creation of the computer. We are by nature inquisitive and creative creatures.

Why are so few people able to capitalize on their inventions? Procrastination is a huge factor. We put off until tomorrow things that we're not comfortable doing. Some people are just waiting for the "right time" and that time never comes. The answer is to get off your butt and do it now.

Also, many people like to avoid risk. A few people are "river boat gamblers," but most would prefer to hold their cards tightly. No large reward ever happens without taking a large risk. Inventing and profiting from your invention is risk taking. There's no way to get around it. There is definite risk of financial loss. No one can make the decision of how much to invest in your idea but you. This book will help you make smart financial decisions!

Most of us prefer to avoid rejection and ridicule. If you are serious about inventing and if you want to make inventing a business, you will receive ridicule and rejection. A faint heart never wins in the invention business. Some of your ideas may be great, but some are probably ridiculous.

My philosophy is that there is no such thing as a bad idea. A lousy idea can spawn a great invention. However, you will be rejected and laughed at from time to time. Get over it. Get a thick skin and move on!

Finally, there are the "one idea wonders." These are people who have one idea and they are convinced that their idea is the best thing that ever happened to the human race. Maybe they are right. This has happened and many people are rich now because of that one great idea.

However, the odds that your first "great idea" is the one to make it all happen for you are slim. If you came up with one super idea, then you can come up with hundreds of other ideas. The more ideas you create the greater your chance of success.

Don't limit yourself to "one and done"! Don't put all your eggs in a one idea basket. Fill that basket up with dozens of ideas, in order to maximize your chance for success.

This book will give you tips on how to keep the creative juices flowing. There are techniques used by professional inventors to help them to continually generate new ideas.

CHAPTER THREE

YOU'VE GOT THE BEST IDEA SINCE THE INVENTION OF THE "RUBBER GRAPE!"

So, you've got this great idea and you're convinced the world needs this idea. And you think you can make millions on it. What do you do now?

EVALUATING YOUR IDEA

I. ORIGINALITY
The very first thing you need to do is check the internet. Even though you haven't seen your idea doesn't mean it hasn't been done. When I first started inventing we didn't have the advantage of having the world at our fingertips on demand. There is no excuse to spend time and money re-inventing a concept that's already out there. As you check the internet, be thorough. Use as many descriptive words as you can think of to bring up any similar ideas. If you can't find your idea, then you're in business... you're on your way.

II. NEED
Is there a need for your product idea? You should take a

sober and unbiased look at your product and answer this question as truthfully and factually as possible.

III. <u>IS YOUR PRODUCT IDEA COMMERCIALLY VIABLE?</u>
One way to evaluate your idea is to consider where it would end up in a large store like Walmart. If you can see your product at Target or Costco, then you have an idea with broad commercial appeal. You could potentially make a lot of money on this product. However, if your product has a smaller "niche" market, then the potential of your idea is reduced considerably.

For example, if your idea helps an electrician in a very specific area, it is limited 1) to electricians and 2) to electricians who see a need to spend money for one very specific task.

However, if you have the next greatest dog toy, then it could be in Walmart and purchased by any of the millions of dog owners.

Obviously, the risk reward is greater for the electrical item. But don't underestimate a niche item. It could generate a small but nice income for years and years. You need to consider the other factors mentioned in this chapter, to properly evaluate whether to proceed.

IV. <u>IS YOUR IDEA PROTECTABLE?</u>

You can make a preliminary evaluation of your idea to determine if it is protectable. If you've determined your idea is original, then you have a shot at a utility patent or a design patent. If you have significantly improved an existing item, you also may be able to get a patent. This is where it helps to have established a relationship with a patent attorney. Some people advise that you can get your own patent. Throughout your journey in this "wacky world" you will get good directions and bad directions. You will get good advice and bad advice. Doing patent work yourself is bad advice. Unless you are an "attorney/engineer wanna-be" you're totally fooling yourself if you think doing patent work is easy.

The dirty little secret is that most attorneys won't touch patent work with a ten-foot pole. They're afraid of patent work because it is so detailed and specialized that it takes years to learn and it requires you to have the skills of an engineer as well as an attorney.

Anyone can write a patent but few people can write a good and complete and effective patent to give you maximum protection for your idea.

You can still succeed with an idea if it's not protectable but it is infinitely more difficult. Companies will shy away from working with you. Unscrupulous individuals and companies will "knock you off" and steal your idea. Foreign companies especially from China/Hong Kong would love to undercut the cost of manufacturing your great idea and sell it for less!

V. IS YOUR IDEA COST-EFFECTIVE TO MANUFACTURE?

There are many great ideas that are just too damn expensive to be practical. As the creator of an idea you will invariably be asked how much will this cost to manufacture. Of course, the answer to this question is often a detailed and complicated one.

Most inventors don't have the manufacturing background to answer this question. Here are a number of the questions that you will need to answer:

1. Will you make it in the USA or overseas? Obviously, the cost to manufacture overseas will usually be less – and usually significantly less. However, more and more

manufacturing is coming back to the USA. Why? Because costs in China and other countries are increasing. Labor costs are up and raw materials costs such as plastic are going through the roof.

2. <u>What is needed to manufacture your concept?</u> Do you need a mold or multiple molds? What will the molds cost? How many paint applications will be needed? How much engineering has to be done? Do you already have a good prototype or are you trying to sell off of computer generated drawings? The cost of molds is extremely expensive. The art of designing your idea to require the fewest molds possible can save thousands.

3. <u>How many do you anticipate manufacturing in the first year?</u> If you sell your concept to a large manufacturer, you don't have to answer this question. However, if you're doing self-manufacturing then this question is extremely important. Here is a secret tip for self-manufacturers: always underestimate what you can sell this first year. Inventors who manufacture their own products are almost always overly optimistic about the quantity they can sell in the first year. There are thousands of game manufacturers who think they have the hottest new game of the century and pay for a first run of thousands of copies. Why? The more you run the cheaper they are. Unless you don't sell them! There are thousands of garages around the country being used as storage places for all these "hot games" that fizzled after a few sales.

It's even worse if you've paid for an expensive mold that will last for years vs. an inexpensive short run mold. If sales fizzle, you end up sitting on a useless mold or worse yet someone in China will end up with that mold and use it or melt it down.

Do your best to calculate the cost of manufacture before you pitch your concept. And if you are going to self-manufacture think small or better yet get some orders before you spend a lot of manufacturing dollars.

CHAPTER FOUR

TO PROTOTYPE OR NOT TO PROTOYPE – THAT IS THE QUESTION!

You have this great idea. You start with a few rough sketches on a pad of paper. You resketch your idea over and over again until you get your great new invention to a point where not only you – but other people – can see the "magic" in your concept. But where do you go from there?

You check around and find you can have a prototype made but it will cost you big bucks - $5,000 to $10,000 – more? What do you do?

It is immeasurably easier to sell a concept if you have a working prototype! Your prototype proves your concept. It proves to a third party that your idea works.

If a picture is worth 1,000 words then a prototype must be worth 100,000 words. Hands on sells your product. While it is possible to sell a concept from a rendering, it seldom happens. It is the exception.

People are tactile oriented. They like to feel things. They like to see things actually work, even if they're not from Missouri, the "show me" state.

Also, a bad prototype can kill an idea. Maybe you've proven you have a bad idea. Maybe it's too expensive to manufacture. Maybe you've invented a new Frisbee, but it doesn't fly! If so, you've just saved yourself a lot of money by shelving your project then and there.

Or maybe you just have a lousy prototype. You may need two, three or four prototype revisions to get it just right... to get a proto that really works the way you planned it and to prove your concept.

Prototypes don't always have to be pretty. Any manufacturer worth his salt can see past an ugly proto if the proto proves the concept, i.e. proves you have a great idea.

How do you find a good prototype maker for your product concept? Just like everything else, the answer has changed. The first thing you should do is check the internet. If there is someone in your area who has the capabilities, then you need to do a face-to-face with this person. Ask questions! What can he do? What has he done? What will the cost be? What is the time frame?

Of course, the type of prototype needs varies greatly. Will it require machining? Can it be made with a 3D printer? 3D printers have revolutionized the ability to make working prototypes at a reasonable cost.

Your prototype may require sewing or pattern making. Find a retail outlet in your area that sells supplies to people who work with fabrics. They will likely know someone who will work for an hourly rate or better yet an agreed upon amount.

Another source for a prototype maker is your local technical college. Check with instructors who work in the subject area of your concept. They may know of a talented student or the instructors themselves may be interested in doing your prototype work on the side.

My first two model makers were both instructors for the model making program at a local technical college. They both started making prototypes for me part-time and one decided to work for me full-time. This person actually made the working prototypes and packaging for the first Micro Machines which helped me sell the concept to Galoob Toys.

In addition, the more you interact with suppliers and small shop manufacturers in your area, the greater the likelihood that you will find a competent prototype maker.

Here's some more good advice: As I said before, make sure you negotiate the price for your prototype or at least a not to exceed price. Then make sure you put it in writing! People have a tendency to forget the details of oral agreements, especially if forgetfulness is to their advantage.

Also have these prototype makers sign a good CDA (Confidential Disclosure Agreement). Chapter Eight of this book discusses CDA's and gives examples of good and bad disclosure agreements.

Finally make it clear that you are hiring the model maker for a fee and that all rights in the concept belong to you.

A word of caution. You may be tempted to give the prototype maker a piece of the action. This may be in lieu of or in addition to a cash payment. Resist this temptation if you can. There are plenty of good proto makers that will work for a fee rather than getting a percentage.

However, there are situations where the proto maker adds an important element or elements to your idea to actually make your concept work. In this case, you might consider a cut to the designer.

However, you will need a formal agreement or letter of understanding signed by both parties to clarify your change of relationship.

Also, a lot of people who come to me have already spent every dollar they have on a patent or with a scam company and just don't have any money to put into a prototype. Each case is unique and has to be evaluated on its merits.

One way or another, it helps to have a working proto to effectively market your concept to a manufacturer. This proto will also help your patent attorney protect the details of your idea.

Chapter Five

HOW DO I "KEVLAR" MY IDEA?

Unfortunately, the patent law has recently changed from protecting the first person who conceives an invention to protecting the first person who patents that invention.

In the past, you could document the day you conceived an invention with a drawing and witnesses and have some protection for your idea. However, in 2012 the patent law was revised. Now it's a first come, first protected law. Instead of first to invent, it is first to file. If you don't file first, you are not protected.

Therefore, protecting your idea will cost you money. If your idea is protectable with a design patent or a mechanical patent you can apply for a provisional patent. This gives you temporary protection for one year for a minimal amount of money. Thus, you have a year to determine if you want to pay the much higher cost of a regular patent application.

Provisional patents will cost you $1,000 and more. Mechanical patents will usually cost you $7,000 to $10,000 to "the sky is the limit." Design patents by their nature are less involved and of course give you less protection and the cost is therefore less.

For written works or works of art or design you can also obtain trademark protection at a fraction of the cost.

Copyright protection is also available for printed material that is disseminated in the public domain.

My words of wisdom for anyone obtaining patent protection is be aware that your initial fee is not the end of your cost.

Our government in its wisdom and search for ongoing revenue has decided that you as a patent owner should pay ongoing maintenance fees. And guess what. If you don't pay the expensive maintenance fee you lose your rights to protection under your already expensive patent!

Chapter Six

DO I NEED PROFESSIONAL HELP?

When people get an idea for an invention and they decide in their mind it's a good idea, what do they do?

First of all, they usually tell their family members or close friends. If it's a toy idea they may tell their children. If it's a household idea they'll share it with their spouse. If it's a work place idea they'll share it with a co-worker. Hint. The more you share your idea the better chance you will lose your confidentiality protection. Keep it in the family.

Most of the time the people you share your idea with will confirm your excitement for what you believe is a great idea. At this point in time you must decide whether to have your concept legally protected. If you go on the internet you will be overwhelmed with companies that want to help you. Beware. These companies have honed their pitches to prospective clients to make you think that all they want to do is help you. Actually, most of them are designed to get you to pay as much as possible all while confirming your feeling that you have invented the next great product that people can't live without.

Here's a secret you should know about attorneys. You can talk to an attorney about your product without paying anything. Most attorneys will allow you time to consult with them about taking your case without charging you.

So, here's your first step. You check the internet for patent attorneys in your area. This may be a mile away or 100 miles away from you. You then call to schedule an appointment with that attorney to discuss your invention and to get an estimate of the cost to you, i.e. all the costs including filing and maintenance fees. You also confirm that there will be no charge for this consultation time in order to give you a cost estimate.

After you've done this you do it again until you've scheduled at least three appointments with patent attorneys in your area.

The advantage of talking with patent attorneys is significant. First of all, non-patent attorneys usually are useless when it comes to inventions. They shy away from anything "patent" like the plague. They don't understand patent law and they don't want to.

Secondly patent attorneys deal with inventions and creative people for a living. They know what to do and how to do it in the most effective way to protect you and your idea. Some of them may be expensive but at least most are qualified through experience.

Third, many patent attorneys are also engineers or have an engineering background. So, when you talk to one, two or three patent attorneys you are also getting the benefit of their engineering background as they evaluate your idea. Is it cost practical? How will you make it? Does it solve a problem?

Finally, patent attorneys work with other professionals such as designers, prototype makers, etc. who you may want to enlist to assist you. Patent attorneys are a great source of information and contacts for you.

CHAPTER SEVEN

HIRE AN ATTORNEY OR SWIM WITH THE SHARKS

After you've talked with at least two patent attorneys – three is better – then the decision is to hire or not to hire.

There is a saying that a lawyer who represents himself has a fool for a client. A non-lawyer who represents himself is a fool who is fooling himself.

As I said before patent law is an area that is too specialized for the average attorney. If that's the case why do you as a layman think you can do it yourself? But I guess you don't know what you don't know and some people would rather take their chances.

Money is usually an issue. That's why you must pin your prospective attorney down and you must get a firm price in writing. If an attorney won't do this move on to a different attorney.

Here's a general rule you can use in any situation when you are hiring someone to do a job for you. "It is human nature for people to underestimate how long it will take and what the cost will be in order to accomplish a given task." Therefore, it is essential for you to get a signed estimate in writing before you agree to hire a person or company for any task.

Here's another little secret. You can negotiate with attorneys. Never be afraid to ask for a lower price.

Do not let an attorney compare themselves to a doctor. Example: "I'm like a surgeon. Would you expect a doctor to give you a firm price when he doesn't know what he'll find when he "opens up" this case?" If you hear this, move on to another attorney, because you are definitely going to be overcharged.

There are some attorneys who are very conservative and fair with their billing. They are also fast and efficient in what they do. This is how your first interview with them should go. Direct and to the point, i.e. fast and efficient with all of the information you need to make a decision in a short period of time.

However, there are some attorneys who like to see how much they can bill out. They might actually bill out 12 hours of work for an 8-hour day. Or they might double or triple bill. Example. When they draft an agreement, they bill out one hour for their time and two hours for the agreement, i.e. three hours charged for one hour of work.

If an attorney talks too much and tries to impress you too much, beware. Also remember attorneys from large law firms in expensive offices in high rent areas have to charge more to justify their existence. Attorneys from smaller firms or sole practitioners have less overhead to deal with. That's why you take the time to interview three different patent attorneys before selecting the right one for your situation.

How do you know you are selecting the right attorney? Again, check the internet for any ratings or other information about the firm or the attorney. Ask the attorney for references and then make sure you check them. Also, if you can get a referral from a friend or acquaintance or from another attorney or professional, there's nothing like first-hand experience.

Finally, you could test the attorney you select with a small project to see if the result is competent and the charge is reasonable.

CHAPTER EIGHT

DEALING WITH BIG CORPORATIONS – THE 600 POUND GORILLAS

You have taken your exciting new idea from pencil scratchings to an actual working prototype. You've protected your idea, your concept, your prototype by commencing a patent application process with a competent attorney.

Now you are ready to show your product to the world. But who and where and how? First you must decide where your product fits from a retail perspective. Where do you see it being sold? Will it be in one of the big box stores, like Walmart or Target? Or is it a specialty type of item that will be sold to a select group of people?

Here's a hint. Go to the store where similar type products are being sold and look at the other products. On the boxes containing these products will be the names of the manufacturers. Make a list of all the manufacturers of similar products. These companies are all potential buyers of your excellent new product!

Now it's up to you to do the leg work... figuratively. You need to review each of these company's websites. Find out if they will consider receiving concept ideas. Who in their company reviews ideas? Then you need to find out if they will sign your Confidential Disclosure Agreement (CDA) form. More likely they will have their own CDA for you to sign.

Here's another piece of advice: Companies are much more likely to review your product if you have a prototype and if you have a patent. In fact, some companies will not review your concept unless you have a patent or have applied for a patent.

If you have a patent, a CDA is less important. If you don't have a patent a CDA is essential. Your right to patent ends one year after you make a public disclosure. If you show your concept or prototype without a CDA, i.e. a confidential agreement, then you have made a public disclosure.

Companies will usually provide you with one of two types of CDA's. There is the one that protects both the inventor and the company. And there is the one that protects only the company.

If you sign the Non-Confidential Disclosure Agreement (NCDA) that only protects the company then you have made a public disclosure of your concept. Your one-year window to get patent protection starts when you sign an NCDA. Also, this company can take your concept with impunity. But far worse they could actually patent your idea, if they file before you. That is how important it is for you to get a good CDA signed if you are sharing an unprotected concept or prototype.

I have included Exhibit A which is a CDA that an inventor might send to a company prior to disclosing their product. I have also included Exhibits B and C which are examples of the types of Disclosure Agreements companies might send you to sign.

Take some time and read all three of these documents. Note the differences between the forms A and B which both treat your concept as confidential. This is important because it is agreed by both parties that you have not made public disclosure. Note that form C states the opposite. It states that the company is <u>not</u> treating your idea as confidential and that it has <u>no</u> obligation to protect your concept.

Also, companies will frequently limit the amount of time they are obligated to protect the confidentiality of a concept. This will usually be two or three years. Obviously, it's more advantageous to the inventor if no time limit is included. This can be negotiated. Get the longest amount of time you can.

My personal opinion on CDA's for my business is that I am willing to sign the A or B versions in most situations, but I'm very reluctant to sign form C. If you have any doubts, you should consult with an attorney.

CHAPTER NINE

HOLY COW: THEY LIKE MY IDEA!

Eureka! The company has decided they like your idea. "Like" is the operative word here. What most inventors hear is they "love" your idea and are going to buy it. This is like a job applicant who finally gets one interview and thinks in his mind that he has the job. Not so fast. The odds of that particular company actually buying, manufacturing, and selling your product are still less than 50%, maybe a lot less.

So... you have to protect yourself again by putting some limits on what the company does with your prototype when you send it to them.

Over the years I'd hate to think about how many of my prototypes have been returned broken, parts missing or in some cases lost and not returned at all.

You protect yourself with a strong but reasonable Option Agreement. I have included a standard form Option Agreement (Exhibit D) that you may want to use. If the company wants to hold and evaluate your concept, then they should be willing to sign an Option Agreement. Option Agreements, like CDAs, will give the Inventor varying amounts of protection. Companies also will pay inventors varying amounts of money to hold and review from zero to thousands! Hint: always try for the thousands but be reasonable. Also, the amount of the option fee you receive should vary with the length of time they want to hold your prototype.

Remember the longer they hold your concept prior to signing a licensing agreement the longer your idea is off the market and unavailable to other companies and not making you any royalty money.

I have seen instances where companies will try to hold inventor's products off the market for a long period of time, just so the product won't compete with something in their product line or something they plan to come out with in the near future.

A reasonable amount of time for a company to hold and review a product is for 60 to 90 days.

A reasonable option fee is anything you can negotiate. It will depend on the size of the company, the market potential, the cost of your prototype, the projected sales income from this product, and other similar factors.

There are situations where the company refuses to pay an option. If you like this company and want to deal with them anyway, then you can waive the option fee, but still have them sign an option. This will give them a timeframe for review and will protect your prototype in the event it is damaged or lost.

When you give companies a time limit for review, they will frequently want more time. This of course starts the negotiating process all over again. Good luck!

CHAPTER TEN

THE PRODUCT LICENSING AGREEMENT: NEGOTIATING WITH "THE DEVIL"

The toughest thing for most small inventors is to negotiate a Product Licensing Agreement (PLA). Especially when you don't know the difference between a good PLA and a bad PLA.

In most cases the balance of power is definitely with the Company. If the company is a large, multinational manufacturer, you will probably be negotiating with "the devil." Even smaller companies have a huge advantage over the inventor. You may be lucky to find a company that is fair and generous, but remember their obligation is to their shareholders and to the bottom line, not you.

So, you can help balance the playing field by hiring a "competent" attorney to represent you. This person may or may not be the patent attorney you used in order to protect your idea.

Do your due diligence to determine if the attorney you are considering has ever negotiated a PLA. The same questions you would ask when hiring a patent attorney apply when you hire an attorney to negotiate a PLA. If you can't get a firm price, because of the uncertainties of negotiation, then get a range with a "not to exceed" amount. This is the maximum you will pay to finalize this contract. This maximum amount will protect you in the case of a protracted negotiation.

The truth is I could write another whole book on product licensing agreements (PLA's). There are so many factors involved from product definition to tax considerations for the inventor. What I have decided to do is list some of the key things for you to consider in any licensing agreement you are asked to sign.

1. <u>Description of your concepts</u>: As the inventor, you want the description to be as broad as possible. You want the definition of your concept to include future extensions, additions, and variations of your idea. The company on the other hand may want to limit your royalties by defining your concept very narrowly and with very much detail.

    As the inventor of the concept you have the right to royalties from your concept and from whatever related concepts that arise or are borne from you selling your concept to the company. Be firm on this point and make the definition of your product/concept as broad as possible.

2. <u>Sale of all rights or license for a term:</u> What are you selling? Are you selling all your rights with a reversion back to you if the company doesn't manufacture or stops manufacturing? Or are you licensing your product for a number of years, i.e. for a limited term? Why do I ask? Because there are different tax consequences to you. Depending on your situation and the expert advice

from your certified public accountant (CPA) one situation could result in "ordinary income tax" and one situation in "capital gain tax." And as we all know capital gains tax is usually much less. Don't be afraid to get opinions from more than one CPA. You'll be surprised. The answer may be different depending on how conservative your CPA is!

3. <u>Time frames for manufacturing, marketing and sales:</u> Your PLA should give the target time for the manufacturing, marketing and sales of your product. If the company doesn't meet these dates, then you should have the option to nullify the agreement and have all rights to your product revert to you.

4. <u>Subsidiary or Partnering Companies:</u> What happens if the company you're dealing with decides to have another company that they have an ownership interest in manufacture and sell your product? What happens if the licensor (the company) decides to license your idea to a third party? Your contract should be reasonable and should protect you in all possible scenarios.

5. <u>Planning for the future:</u> It would be nice if we had a crystal ball to know exactly what will happen to your product. Then we could draft the perfect PLA to protect you.

What if the company stops manufacturing your concept for one year? What if the company refuses to pay you? What if they don't pay you in a timely manner? What if they don't pay you what you think you should be getting? What if the company goes bankrupt? What if the company gets sued because of your concept? What is your liability? All of these questions need to be answered in the PLA in order to protect you and your family.

6. <u>What happens to the "molds" used to make your product if your contract with the company is terminated?:</u> Obviously you don't want some other company in China or any place to buy your molds and start making and selling your product without your permission. This means the contract should provide that you get the mold back or that they are destroyed and you need proof of this.

7. <u>Does your PLA provide that you are selling all your rights worldwide or just in the USA?:</u> Most companies will want the international rights but some may have no intention to sell in other countries. If you are selling international rights, you must put marketing and sales timelines in your contracts. If the company doesn't meet these timelines, then the international rights revert to you. This is the same if the company doesn't meet its USA marketing and sales timelines, then all rights could revert to you. However, you need to be reasonable and work with the company if they are showing good faith.

8. <u>Does your PLA include penalties for failure to make payments and/or for late payments?:</u> There should be a significant penalty for the company making late payments. This will ensure that you get paid on time. Also, if payments are missed because of accounting errors or intentional deletions then there should be an extra penalty included, which should help pay for your audit expenses.

9. <u>Have you included the right to a periodic audit of the company to verify sales and royalty payments?:</u> You would be surprised how often that the person calculating the royalty payment does not have a clear understanding of what the PLA requires. You'll never know this unless you audit. Of course, the expense of an audit never makes sense unless the royalty amounts are large enough to justify an audit. A good contract will

allow an audit up to once a year at a convenient time for the company. Surprisingly most inventors, even the most successful ones, never audit their companies. Why, I don't know. However, most audits will reveal an underpayment of royalties in one form or another.

10. <u>Is the company required to provide you with samples of each product you receive royalties on under your contract?:</u> This may seem like a minor point compared to the others. However, inventors like to show off and even give away samples of their creative genius.

If your contract requires that you receive samples then the person in distribution will comply. This is usually not the person you negotiated the PLA with. That's why it's nice to have it in writing.

In the past, I've negotiated from one to three dozen samples of each product under the PLA. Companies usually only want to agree to half a dozen.

Also, companies will forget to send you samples unless you ask. So, ask.

Contract negotiation can be very frustrating, especially for a novice. Make sure you are ending up with a PLA that has some solid protection for you.

There is usually a "push and pull" in negotiating a fair agreement. Frequently a fair agreement is one that neither party is 100% happy with. Be firm but be reasonable.

I have included a sample PLA for you to review as a guideline and for comparison purposes (Exhibit E). However, I strongly recommend that you work with an experienced attorney in finalizing any PLA that you sign.

Once you've signed on the dotted line, you have taken a leap of faith. You have put your trust into the license and in the company. You kiss "your baby" goodbye and hope they treat her right... and you!

CHAPTER ELEVEN

TRACKING YOUR PRODUCT...
STAY AWAKE AND KEEP YOUR EYES OPEN

After the company has your product it will take them awhile to get your concept to market. If you put target dates in your PLA then you have some idea when you'll see your idea in a store or online. If not, call the company and ask them.

As the inventor, you need to develop contact people within the company. These are people who can give you information on the status of your product. This could start with projections the company has for manufacturing and sales, from someone in marketing. It could also be someone in their accounting department who can give you the actual sales report. This may be the person who is responsible for cutting your quarterly royalty check.

Hopefully your product will hit the shelves within a year after you signed the PLA. You will then start receiving royalty checks every three months and all will be well with "your baby."

How do you know if the royalty check amount is right? Do they give you a lot of detail with your statement? Or is there little or almost no detail? Has the company made certain deductions for shipping, for marketing, and sales promotions, or a number of other expenses they have in selling your product? Are all of these deductions contemplated in your PLA?

Did you agree to a blanket deduction for all of these expenses not to exceed 4%, 6%, or 8%? You need to look at your royalty statement carefully and ask for more information if you need it. I've found that most inventors are afraid to ask questions or they just don't like to be bothered with financial or contractual details.

Inventors for the most part by their nature are right brained creative individuals. Attorneys and accountants tend to be left-brained analyticals. That's why you need them!

Here's a hint: if your product is on the shelves of the big box stores like Target, Walmart, or Toys R Us, you should be receiving a large royalty check. If "your baby" has ended up in specialty stores your check will be much smaller. Your royalty statement should detail what stores your product is in. Smaller stores may show up listed as a buyer group for a number of stores.

At some point in the evolution of your product you may want to audit the company. There is no way to know if you are getting paid accurately unless you audit. Inventors hate to audit companies for a number of reasons:

1. They are afraid to antagonize the company. Even though your PLA says you have the right to audit, it costs the company time and money. You don't want to appear to be the untrusting bad guy.

2. It costs a lot of money to audit. First you have to find a competent accountant. Auditing is another of those specialties that most CPAs aren't qualified for. Next you have to negotiate a fee. Use the same strategy as already outlined in this book for attorneys.

3. Right brained inventors don't like to get involved with left-brained activities like audits. They would rather try to move on to the next invention. I am amazed at how few toy inventors have ever conducted an audit of toy companies they deal with, even when the royalty numbers are huge. Do a cost/benefit analysis. Are your royalty checks large enough to justify the cost?

I have never conducted an audit where I didn't find that the company had made mistakes in calculating my royalty checks. And in every case the mistakes worked to the company's advantage. Every audit I have conducted has resulted in royalty payments that were due to me. Sometimes large sums of money.

    Here are a few of the things that an audit can discover:

1) Miscalculations of royalties

2) Intentional deletion of sales from royalty figures

3) Failure to pay on mixed product, i.e. your product sold with another product

4) Failure to pay on accessories to your product

5) Failure to pay on some or all international sales

6) Failure to pay on sub-licenses of your product

7) Miscalculation of allowable deductions

8) Credits received by the Company in lieu of payments for your product, i.e. no royalties paid

These are just some of the calculation errors and royalty omissions, unintentional or intentional, that might be discovered as a result of an audit.

CHAPTER TWELVE

THE 9 BIGGEST MISTAKES YOU CAN MAKE AS AN INVENTOR

As you can see from the previous chapters there are many things to consider in trying to get your product to market. It's easy to make mistakes along the way. Here are what I consider to be the 9 biggest mistakes inventors make.

1) <u>Disclosing your idea or concept to the world.</u> When you make any kind of public disclosure of your idea you may forfeit your right to protect that idea. Keep your idea close to the vest. And know what the patent law says about disclosures.

2) <u>Not having a decent prototype made.</u> Prospective buyers need to see your concept in a 3D form. You only have a few seconds to convince a prospective buyer. Use that limited amount of time to prove your concept and impress the buyer.

3) <u>Trusting companies to be fair and honest in their dealings with you.</u> Company executives owe their loyalty to their shareholders and to themselves. You are at the bottom of the list. Many companies will treat you fairly. You just don't

know which ones will and which ones won't. Don't assume anything.

4) <u>Not using a confidentiality agreement (CDA).</u> When you disclose your idea to any third party, especially a prospective buyer, have that person sign an agreement to keep your concept confidential. Remember any <u>non-confidential disclosure (NCDA)</u> tolls the one-year period of time you have to get protection for your idea. It also puts your concept at risk to be stolen.

5) <u>Not protecting a great concept.</u> First you need to objectively evaluate your idea. This book will help you do just that. If you are convinced that you have a great idea, then hire a professional to help you protect that idea. Hire an attorney who is experienced in the area of intellectual property protection.

6) <u>Giving up after a few rejections</u>. You are rejected three times and you decide to give up. Faint hearts never won the fair lady. A faint heart won't sell your great concept. You just have to find the right company for your invention. Rejection is a huge part of this business. If you can't take rejection you are in the wrong field.

7) <u>Not using an attorney for a product licensing agreement (PLA).</u> Contracts are by their nature detailed and complicated. The language used is frequently backed up by case law or statutory law that determines your rights and the rights of the company you're dealing with. Get help. Get a qualified attorney to assist you. Get a good contract. Make sure your contract defines your concept as broadly as possible and protects your rights as the inventor.

8) <u>Failure to accurately evaluate your concept.</u> This book tells you how to look at your idea from different angles. What if no manufacturer will buy it? Can you prove it hasn't already

been done or that you are not violating a pre-existing patent? So, what if it's a great idea if consumers won't buy it?

9) <u>Not knowing when to throw in the towel on your idea.</u> Sometimes you've answered all the preliminary evaluation questions correctly, but no one wants your product. You've been rejected 20 times but you are sure your product is still great. Look at all your rejections. Consider the reasons people give to you. Maybe they have something. Something you have missed in the excitement of the invention process. Shelve your idea for a while and bring it out again in a year or two. Maybe the time is not right. Maybe it never will be. Know when to cut your losses and move on to the next great idea!

CHAPTER THIRTEEN

ARE YOU CRAZY ENOUGH TO DO IT YOURSELF?

You've talked to a bunch of manufacturers and they've all blown you off. Or you just like to cut out all the middlemen. Or your product just isn't that hard to manufacture and you're looking for a new career and you like to work 12 to 16 hours a day. Yes, you've decided to self-manufacture and self-market.

The reality of self-manufacturing is that you will have to wear many hats. You will be responsible for manufacturing. You may be the one wearing the hard hat, holding the clipboard, or stirring the pot. Maybe you'll do it in your garage. Maybe you contract out some or all of your production.

As a manufacturer, you have to keep your costs under control. If you've decided to manufacture in the USA, you are definitely being patriotic. The problem is 9 out of 10 times it will cost you more to make in the USA. Sometimes a lot more.

The advantage is control over the product. This usually means better quality. One of your main tasks is to provide a quality product while keeping the cost down. Manufacturing costs have killed many good products.

You have to get firm bids from multiple manufacturers. You need to do your due diligence with whoever you are going to work with. Get the names of other customers of theirs and find out how they are to work with. Do they communicate well? Are they on time with their production? Do they hit you with cost overruns and surcharges?

You are wearing the "big boys or girls hard hat," so the buck stops with you! Any time you have left over will be consumed with marketing and sales. A good idea would be for you to start out with a plan. What is your marketing strategy? How will you get in front of buyers? Are you going directly to consumers or are you going to go through distributors?

The advent of internet marketing has enabled you to sell your product directly to millions of prospective buyers. There are numerous companies that have become overnight success stories because they had a product they could sell through their website. Some sell just enough to keep going and some sell a lot. For many inventors, internet marketing has opened up a whole new world of possibilities. The internet has ushered in a new and exciting "grand age of inventing."

How to get started with internet marketing and how to do it all correctly would take up half this book. There are numerous self-help books on the internet that can give you the basics, if you want to pursue internet sales.

Sales projections are a challenge for all companies and even more so for a self-manufacturer with limited capital. You have to have enough product to satisfy the consumer demands, but you don't want so much inventory on hand that it weighs down your whole business, such that you have all your working capital tied up in inventory. This is inventory that you've paid for which is now sitting in a warehouse that you're also paying for. Be smart, run lean but have enough on hand to keep "your machine" running.

If you are self-manufacturing here are some other considerations:

<u>What type of business entity are you?</u> Are you an individual, partnership, Corporation, or an LLC? Consult an attorney or a CPA for advice on what entity works best for you and your situation.

<u>Do you have good insurance?</u> You need business insurance that fits your product and your operation. You want to strike it rich, you don't want to fall down the mine. As a business person, you have various forms of potential liability, from employee liability to products liability.

Fortunately, there are many qualified and reputable insurance companies ready and willing to assist you and charge you for it. Get bids and look at the details. Oh yes, and read the insurance contract!

<u>Make sure you pay your taxes.</u> As a self-manufacturing small businessman, you will be liable for various taxes, from employee taxes to personal property taxes to income taxes. Make sure you have a relationship with a good CPA who will lead you through the myriad of tax regulations that apply to you. And remember always pay your taxes! If you don't, there are consequences!

Self-manufacturing offers up many challenges to an inventor/entrepreneur. However, if you have the stamina and personality for it, you can find it both personally satisfying and financially rewarding.

CHAPTER FOURTEEN

THE 9 BEST THINGS I'VE DONE AS AN INVENTOR

1) <u>Kept a written record of all my crazy ideas</u>

    Whether it's a notebook or a legal pad or scraps of paper, memorialize your ideas no matter how crazy by putting your ideas in writing to preserve them. Until they changed the patent law you could actually have some protection for your great ideas by having your handwritten design signed, dated, and witnessed. This won't give you much protection anymore. But it still isn't a bad procedure for those ideas that you think could have wings.

    I like to write down ideas as soon as I have them even if it means getting out of bed at midnight to do it. There's nothing worse than trying to remember that great idea you had at 2 a.m., but you just can't recall what it was. Keep a notepad next to your bed.

    Some of my best ideas have come when I'm in the "twilight zone" halfway between awake and sleeping. Better be safe than sorry. Write it down!

2) Made Working Prototypes

You have such a short time to impress a prospective buyer. Prove your concept with a prototype. Words are good. Pictures are better. A working model is by far the best!

3) Legally Protected concepts when feasible

Not every idea needs to be protected with a patent. You need to do some serious evaluation and soul searching. Legal protection costs a lot of money. But if you have that special concept and you've done your due diligence, then by all means protect "your baby."

4) Used professionals

My background as an attorney is in trial law, family law and real estate. Throughout my years as an inventor I've gained some expertise on patent law or as they call it "idea protection law." However, I still use patent attorneys to protect my concept. Why? Because it's a very specialized area of law and the few attorneys doing this work are unique. Most have some kind of a background in engineering. Many have dual degrees in Law and Engineering.

I also would suggest that novices get help from an attorney for the Product Licensing Agreement (PLA). Even negotiating the terms of the PLA is an art form that should be done by someone who knows what they are doing.

5) Required disclosure agreements

A fair confidential disclosure agreement (CDA) protects you and the company. In Chapter Eight I have explained the different types of CDA's and I have included samples as exhibits for you to use. If a company insists on using a

CDA that gives you no protection, this is probably a company you don't want to deal with. Beware of these companies.

6) <u>Tried to get as strong a licensing agreement as possible</u>

You or your attorney may have to negotiate your way to a strong PLA. You never know how good a deal you can get if you don't ask for it. But be reasonable. Find out what the industry standards are and be reasonable. For example, toy companies will traditionally pay higher royalties than gift companies. Work with them.

Also submit your licensing agreement to the company first. Smaller companies may like to avoid the cost of an attorney preparing a PLA and they may be willing to negotiate from a PLA that you submit. This is a plus for you. And remember, define your own concept broadly to include future extensions, additions and variations of your idea.

7) <u>Attempted to deal with honorable and reputable companies</u>

After you've been in the business for a while you will find out who "the good guys" are and who you should avoid. I avoid dealing with companies that have a bad inventor reputation.

The old saying is cheat me the first time, shame on you, cheat me the second time, shame on me! Companies get a reputation based on their previous dealings with inventors. Ask questions and check with other people.

Develop a working relationship with people in the companies you want to deal with. These individuals can do you a lot of favors or they can lock the door on you. Good relationships are invaluable.

8) Developed strong industry ties

Know what's going on, what's current in the industry you are inventing in, be it electronics, gifts, consumer products or toys. You can go to trade shows and accumulate knowledge and contacts that way. Or you can subscribe to Industry Publications to get the latest news and information.

You certainly don't want to spend time inventing a product that has just been previewed at a consumer-products show. Know what is going on in the industry you are inventing in.

9) Audited companies when justified

Surprisingly few inventors will ever audit the companies that are selling their products. What are the reasons? They don't want to offend the company. They don't know who to use for an audit. There just aren't enough royalties involved to justify an audit.

However, every audit I've conducted has shown errors that resulted in the underpayment of royalties. Every audit I've conducted has been worthwhile. So, make sure you have a good audit paragraph in your PLA.

In summary, never assume anything. This is your idea – your product. No one else is motivated to protect your interests.

Chapter Fifteen

SUMMARY

As an inventor, you have just started or you have been on a long sometimes perilous journey. There are many words that could be used to describe what inventors go through:

Exhilarating. Frustrating. Exciting. Depressing. Challenging. Confusing. Expensive. Tiring. Fulfilling. Rewarding. Educating. Invigorating. Hopeful. Sad. Happy. Angry. Wonderful.

These are just a few of the emotions you go through on this long and perilous adventure. The highs and the lows. The challenges you are faced with to get your idea to market and to the consumer.

You will win some and probably lose a lot. Inventing is not for the faint of heart. You need to be strong and steadfast in promoting "your baby." You also need to be realistic and practical. There is nothing like inventing. There is no greater feeling than when you really have conceived a new breakthrough idea. Whether it's the "rubber grape" or a cure for the common cold. Whether your idea is earthshaking or just a fun new idea to make life easier for people. There is nothing quite like being an inventor. Good luck and God bless.

Read on for a special preview of

AM I CRAZY? NO. I'M A TOY INVENTOR.

By Clemens V. Hedeen, Jr.
Inventor/Attorney/Author

Available April, 2018

SPECIAL PREVIEW OF

AM I CRAZY? NO. I'M A TOY INVENTOR.

What the hell am I doing standing in a broom closet? This is New York City, home of the sleek luxurious high rise mega offices and I'm standing in a broom closet.

Actually, by broom closet standards, this one isn't bad. For one thing, it's fairly roomy about four feet by eight. In one corner is the standard grimy yellow bucket with the standard-duty grey mop. The mop handle is resting next to the handle of a large push broom, next to the handle of a large push mop.

I'm standing at the other end of the closet. Why am I standing in a broom closet? I graduated at the top of my law school class at that famous party school in Madison, Wisconsin. Well actually I was at the top of my class the first year, but boredom was my excuse for not attending classes and I settled in at the middle of the pack.

After graduation, I found my niche in Milwaukee as a trial attorney for the best negligence defense firm in Wisconsin. Now I'm standing in a broom closet on the 11th floor of 200 Fifth Avenue in New York City.

Next to me is a large steamer travel case which I selected after a lengthy search and which I modified by adding caster wheels to the bottom so that it can be easily maneuvered into elevators, through doorways, into taxi cabs (if I could get one to stop) and into airport check-ins (no size or weight limitations then).

I have an 11:00 a.m. appointment to meet the Director of Product Development at a large corporation on the 11th Floor of 200 Fifth Avenue.

I've been waiting in this broom closet for 10 minutes now. I'm supposed to meet a guy in here. How long should I wait? I usually revert to the college mythical requirement... like 15 minutes for a full professor... 10 minutes for an assistant professor and 5 minutes for a TA... or something like that.

Any way the guy I'm waiting for is someone who is probably like an AP (assistant professor) in the corporate world... so 10 minutes isn't unreasonable.

Actually, while I'm waiting I've opened my steamer case and removed a few objects to have them accessible.

The door finally opens and in waltzes a corporate dandy. Gucci shoes... suit... dressed to the nines... in the broom closet.

A quick introduction, the usual greetings, and then a statement that he only has a few minutes to spend in this broom closet with me and finally his question, "what have you got to show me?"

So, I guess I don't rate the plush glitzy office – but come on now a broom closet? Isn't Mr. Dandy afraid that he might step in that grimy yellow bucket next to me with one of his (shined up) super polished Guccis? I guess not. He's done this before.

After all I'm no longer a hot shot young attorney… no… I'm a lowly toy inventor. Toy inventor in the world of the large toy Megopolis is someone you can walk all over, make wait for hours, cancel appointments with or if necessary meet with in a broom closet.

Actually, I should feel privileged. It's not easy to gain access to the world of toys… as an inventor. It's a closed society of about 200 individuals and small companies. Back then when I was meeting with Mr. Dandy there were twice as many of us, but now the economic reality of a dwindling market has taken its toll on toy inventors. Kids have grown up faster and they have graduated to other activities faster, voila… enter the world of video games… computers, etc., etc.

Think back 20 or 30 years ago. Kids played with real toys and table games that you actually played with on a table. They even ran around outside a lot. Now they can sit in a soft chair, play their handheld video games and munch away.

Back to the broom closet. I have four new inventions that I want to show Mr. Dandy. These are the latest and the greatest from the creative mind of myself and my associates… at Fun City USA. I like to put the word "fun" in my company name because that's what I think a good toy should be – fun to play with – again and again.

I'm the creative guy – me an attorney – go figure. My associates can make great 3-D working models. Who would have thought an attorney could be creative.

So, I explain in living color with the help of my usually working prototypes why kids would die to have the toy we've just created… in fact… couldn't live without it.

Unfortunately, my enthusiasm is not shared by Mr. Dandy. The first one is too expensive to produce. The second and third ones aren't on target with their marketing and development plans. The last one he just doesn't get at all... why should he? This guy probably never played with toys in his life!

Anyway, I thank him for the opportunity to degrade myself by imposing my concepts on him in a broom closet and tell him I'm looking forward to meeting him again, next time maybe in a storage room.

We exit from the closet and I find myself in a sea of suits and fast-talking guys from another country – New York/New Jersey. I wonder what Old York and Old Jersey was like?

Obviously, this toy company is doing well. You can tell by how busy their showroom is and it's busy. Unfortunately, any tours of their showroom are reserved for buyers. Inventors just get in the way and they might just say something negative – like holy cow... they stole my toy idea! Toy companies like to make it challenging for inventors... let's make them invent creative toys for us with half their brain tied behind their back. Actually, the idea is we might steal their "new toy" and knock it off. When in reality it's the other way around.

So, I schlep my huge cargo case through this mass of corporate sales asses and make my way to the door stopping at the receptionist candy bowl for a handful on the way out.

The toy building at 200 Fifth Avenue that I'm in has a very unique history. It's actually two buildings connected by a bridge at the ninth floor to 1107 Broadway. These buildings contained more than one million square feet of rental space. Over 600 toy or related companies occupied these buildings.

The elevators at 200 Fifth Avenue are real antiques – beautiful to look at – but they don't always work.

Each elevator is crammed with five or more people than should be allowed and that's probably the reason that sometimes they just stop working... between floors. But don't worry they'll get you out in an hour or two. Hope you're not claustrophobic!

So, I cram into an elevator with my cargo case and head up to the ninth floor, walk across the bridge and then take an elevator up to the 11th floor... Kenner Toys. Not to show concepts because I actually get to preview their new line of toys for this year. AND because our newest toy concept has made it into their line... a toy gun that shoots soft darts... the first one of its kind. **The first Nerf dart gun!**

I'm excited to see just what they've done with our concept. How close is it to our prototype? How are they going to market it? How are the buyers receiving it? Does it look like it will be a hit?

I exit the elevator at the Kenner Floor and I'm again engulfed by a sea of sales fish. Slick guys in their slick suits with slick hair. The sales ladies are up to the competition and are even slicker than the men. There's excitement in the air... new products and new sales. The Kenner staff and their reps are ready to extol the virtues of their new toy line for 1990.

Back then toys were shown in February for inclusion in the retailers Christmas season starting late October. Of course, things have changed now. Manufacturers start showing their toys at the Dallas Toy Fair in October and before for the fall/Christmas season the following year.

The big boys demand more lead time. Toys R Us, Wal-Mart, and Target see the toy lines months before. They receive VIP treatment with personal on-site previews of the manufacturer's toy lines.

Also, the toy manufacturers especially the big ones like Mattel and Hasbro need more lead time. They plan two or three years in advance. Everything takes more time nowadays, planning, models, manufacturing, safety testing, sales testing, and shipping.

Finally, it's our turn to enter the secretive showroom area. Our tour guide is a young corporate guy with the proper Kenner ID on his suit coat. Eight of us are going through at one time. We are a non-sales group – so we're instructed to stay clear and make room for any buyer groups.

Each new product is displayed in a room worthy of a Broadway set design.

The product presentations are made by actors, actresses and models trying to supplement their income while they wait for the "big break."

A glitzy set and a polished presentation about the star of that room – a new line of the latest and greatest toys for the future. Toys that are must-haves "for hungry retailers."

The toy business is the ultimate fashion business. While consumer products companies have products that stay in their line for year after year, the life of a good toy, a successful toy, can be in a line for as little as two or three years.

Thus, the reason that today's toy industry has so many licensed products – Mickey, Sesame Street, Batman, Star Wars, Marvel – and so many branded products – Uno, Slinky, Nerf, and Barbie.

Consumer recognition of a license or a brand adds instant credibility and recognition to a product in a sea of choices – without the very expensive television advertisements.

As our group travels from one mini showroom to another I marvel at the spare no expense attitude – big dollars are involved in the successful launch of each new product and only a handful will be truly successful. But those successful toys will generate millions and in some cases billions of dollars.

Finally, we arrive at "the showroom" for "the" new toy... our new toy. The one and only, **the first ever Nerf dart gun – the Sharp Shooter**. No one had invented a gun that shoots a safe foam dart. Our gun was conceived and engineered at our workshop in Sturgeon Bay, Wisconsin.

We knew we had a great idea and invention but we didn't know how big it would become.

Howard Bollinger the head of research and development at Kenner loved the gun from the start.

Howard was a great product person, a dying breed, a dinosaur in the toy business of today. He could find great products and push them to manufacture. He didn't care if the item was in house (invented at Kenner) or from an inventor. Howard and Kenner were truly the inventor's friend. Working with Howard over a seven-year period of time we sold Kenner and Hasbro 11 different toy concepts from the **first seven Nerf dart guns**[1] to the first Whistling Sound Darts to Furever Friends plush animals. We also sold Hasbro the Nerf Bungee Blaster and the Nerf Double Blast.

In contrast, today most toy companies view inventors as their competition instead of their partners.

The "bean counters" are in charge and are more concerned about how much money they pay out instead of making innovative toy products.

Year after year retailers are looking for the new creative products like **Micro Machines** (also invented by me), Pound Puppy, Cabbage Patch Dolls and Zhu Zhu Pets, but all too often they are stuck with brands and licenses hiding boring products.

Our new **Nerf Sharp Shooter** soft dart gun looks great. The enthusiasm in the room for this new way to shoot a soft projectile with a suction cup on the end that sticks to any smooth surface is obvious. The tall toothy blonde model/actress showing our gun has no problem pulling back the latch mechanism to cock the **Sharp Shooter** and then with a quick pull of the trigger our soft dart flies out of the gun and sticks to the display target with amazing accuracy. Looks like we've got a hit!

Little did we know how big a hit. So big that twenty years later the whole line that this **Sharp Shooter** started is now more than a billion dollars in sales each year.

Of course, Hasbro doesn't want to pay us anymore and after 20 years we're still in court with Hasbro's corporate attorneys trying to argue that they can change the name they put on their latest gun and not pay us a royalty. Such is the lot of the lowly inventor. They love us for new innovative ideas, but they sure do hate to pay out royalties that could hurt their bottom line.

So, we're in court again. It helps that I'm a lawyer, but not that much, because it still costs a lot of money to fight the big boys. Most inventors are "left-brained" and avoid legal fights and courtroom confrontations. Fortunately, I've got enough right brain activity to be dangerous.

The conundrum that most inventors face is when to push your claims against a toy company and when to let it slide, waiting for the next big one – perhaps with that very company.

There used to be more toy inventors and more toy companies. Toy inventors in the 80's and 90's numbered about 500. Now that number is closer to 200. Toy manufacturers grow their business by buying up other toy companies with solid brand names. This is easier and safer than coming up with new innovative products. The "Slinky" brand is a prime example. Doug Ferner and Ray Delvecchio at Poof toys saw the opportunity to purchase the Slinky brand. They had the foresight to realize that this great toy which had been around since 1945 would and could be around indefinitely and that the brand name was underutilized. They were right! Slinky sales and licensing has exploded since they took over the brand.

Another brand name that was underutilized for years was Tonka. Hasbro purchased the Tonka brand and let it sit on the shelves for years. Finally, some small manufacturers licensed the Tonka name and brought Tonka back to life.

So, with fewer manufacturers, when does an inventor risk severing the special relationship that you have with a toy company because you feel you've been screwed?!

Not an easy decision. My formula: does it make financial sense to push the issue? Some toys are great but they will never sell through. It's the ones that have the big potential that you have to protect.

Anyway, I leave Kenner's showroom walking about two feet off the ground and that's not easy, since I am schlepping a steamer case full of toy prototypes.

Tomorrow I have a few more manufacturer appointments, Ideal Toys, Coleco, Just Toys, and Galoob. These are all toy companies that don't exist anymore. Either they got too big for their britches and believed their own marketing or they were so successful they got eaten up by larger and hungrier toy companies, like the Muncher Garbage Truck toy that picks up anything in its way. What a great toy mechanism.

Why they picked New York in February for Toy Fair says a lot about the toy industry. You could write a book about the evolution of the toy industry and its relationship to New York. How manufacturing started and evolved – then moved to China. Why New York was the hub and how the real action now takes place at Walmart, Target and Toys R Us. And how toy inventing started big time in Chicago at the Marvin Glass firm. So, I did.

[1] 1993 - Nerf Sharpshooter – Kenner, 1994 - Nerf Bow and Arrow – Kenner, 1994 - Nerf Sonic Stinger – Hasbro, 1995 - Nerf Crossbow – Hasbro, 1995 - Nerf Sharpshooter II – Hasbro, 1995 - Nerf Max-Force Sawtooth – Hasbro, 1995 - Nerf Max-Force Eagle Eye – Hasbro

EXHIBIT A

CONFIDENTIAL DISCLOSURE AGREEMENT (CDA)

(NAME OF INVENTOR), hereinafter referred to as **INVENTOR** represents that it has developed a concept for the toys and/or game(s) and or consumer product(s), which will be disclosed to **(NAME OF COMPANY),** hereinafter referred to as **COMPANY** pursuant to this agreement. As used in this Confidential Disclosure Agreement, the term "Property" shall mean the concepts and items illustrated in such pictures to be described in future disclosures and shall include any configurational designs, functional designs, processes, inventions, patents, trademarks, copyrights and/or applications in process or prosecution therefore, pertaining thereto. INVENTOR desires to disclose the Property to you for the specific purpose of enabling you to evaluate your interest in distributing the Property under an Agreement with INVENTOR.

The Parties hereto mutually agree therefore as follows:

1. You will maintain in strict confidence and not disclose to any party any disclosure made by INVENTOR of, or relating to, the Property. This obligation of confidentiality shall not, however, apply to information concerning the Property (a) which is or becomes without your fault, known to the public in the form of a printed publication or otherwise part of the public domain, (b) which was known to you prior to your receipt from INVENTOR, as shown by your written records provided, however, that such prior written information is transmitted to INVENTOR, in writing, together with reasonable proof thereof, within thirty (30) days after the date of this Agreement, (c) which is acquired or received by you from a third party which has not derived such information from

INVENTOR under a confidential disclosure agreement which would prohibit such disclosure or (d) data which is specifically released in writing from confidential status by INVENTOR.

2. You shall, in the furtherance of your obligations to retain the information in confidence (i) limit access to the Property to those of your responsible employees reasonably requiring same for the specific purpose aforesaid; and (ii) Require said employees having access to the Property not to divulge the same to any person or concern or use of same without the prior written permission of INVENTOR; and (iii) not use the Property in any way other than for the specific purpose of determining the suitability of your manufacture and selling of products embodying the Property under license from INVENTOR.

3. You shall have no right or license of any kind in the Property so long as it must be held in confidence as provided in this Agreement or any patent rights relating thereto.

4. INVENTOR warrants that it has the right to make the disclosure of the Property to you.

5. It is understood and agreed that the execution of this Agreement is for the sole purpose of protecting the confidential status of the Property and it does not give you any right to manufacture, use and/or sell products embodying the Property. Any such rights are granted by INVENTOR to its licensees only pursuant to written exclusive license agreements.

6. This agreement shall be effective for all products disclosed by INVENTOR to COMPANY from now until this agreement is amended or changed by the written agreement of both parties.

IN WITNESS WHEREOF, the parties hereto have executed this Agreement as of the dates set forth below.

COMPANY   INVENTOR

By: _____   By: _____

Name: _____   Name: _____

Date: _____   Date: _____

EXHIBIT B

CONFIDENTIAL DISCLOSURE AGREEMENT (CDA)

This agreement is entered into between (Name of Company), a company having its head office at ADDRESS, City, State, Zip, Country, and its wholly owned subsidiaries ("COMPANY") and ("INVENTOR"), on the _____ day of _____, 20__.

1. The INVENTOR owns certain confidential information, including but not limited to designs and prototypes, relating to a product concept described as:

   _____

   _____ ("Confidential Information");

2. THE COMPANY is in the business of manufacturing and selling toys and games;

   The Inventor desires to disclose to (Name of Company) the Confidential Information and (Name of Company) desires to review the Confidential Information to evaluate whether to pursue an agreement to license from Inventor. In consideration of the mutual promises made herein, the parties agree as follows:

3. Inventor warrants that it is the sole owner of the Confidential Information and that its disclosure or any grant of rights thereto to (Name of Company) shall not violate the rights of any third party.

4. (Name of Company) accepts the Confidential Information for the sole purpose of evaluating it for license. All information intended by Inventor to be protected under this Agreement shall be in writing and clearly marked as confidential at the time of disclosure. (Name of Company) shall not use the Confidential Information, except for evaluation, and shall not disclose the Confidential Information to any third party without Inventor's consent. (Name of Company) shall take all reasonable steps to maintain confidentiality of the Confidential Information.

5. Inventor recognizes that (Name of Company) may be required to disclose the Confidential Information to members of (Name of Company's) organization for evaluation. (Name of Company) will inform every such member of the obligation to protect the Confidential Information.

6. Inventor acknowledges that (Name of Company) receives numerous submissions that may be similar or identical to the Confidential Information, and the adoption by (Name of Company) of any alternative submission (as opposed to the Confidential Information submitted by Inventor) may be due to market conditions at the time such alternative submission is received. Selection by (Name of Company) of alternative submissions shall be without obligation to Inventor. Where the submission is of a general concept for consideration as part of a (Name of Company) established brand or line extension, Inventor acknowledges that many concepts are considered for line extensions and that any similarity between Inventor's submission and a possible (Name of

Company) created line extension shall not by itself be considered Confidential Information.

7. (Name of Company) shall have no obligation to keep confidential any of the Confidential Information which (Name of Company) can show was known to (Name of Company) prior to disclosure by Inventor; was or becomes known to the public or generally available to the public through no act of (Name of Company) contrary to this Agreement; is or was disclosed by Inventor to any third party without obligation to maintain confidentiality; is received in good faith by (Name of Company) from a third party and is not subject to an obligation of confidentiality owed by that third party to Inventor; or is required to be disclosed in a judicial or governmental proceeding.

8. During and after the term of this Agreement, Inventory shall not disclose to any third party the terms of any proposed or executed license agreement with (Name of Company), nor any information concerning (Name of Company's) products or business.

9. No license under any intellectual property right is either granted or implied by the conveyance of the Confidential Information. Neither this Agreement nor the receipt of the Confidential Information shall constitute or imply any promise, intention or commitment by (Name of Company) to pursue a license from Inventor.

10. The parties do not intend for this Agreement to create an agency or partnership relationship between them. This Agreement represents and expresses the entire Agreement of the parties and supersedes all prior agreements. Any amendment or modification of any provision must be in writing and executed by both parties. The laws of the (STATE) govern this Agreement. All disputes arising out of the covenants hereunder shall be submitted to the exclusive jurisdiction of the (STATE) Courts.

IN WITNESS WHEREOF, the parties have executed this Agreement as of the date first written above.

COMPANY   INVENTOR

By: _____   By: _____

Name: _____   Name: _____

Date: _____   Date: _____

EXHIBIT C

<u>NON-CONFIDENTIAL DISCLOSURE AGREEMENT (NCDA)</u>

You and (Company), which will be referred to as **"COMPANY"** agrees as follows:

1. This Submission Agreement (**"Agreement"**) applies to all materials and information You disclose to COMPANY (**"Submissions"**), including, without limitation, concepts, designs, mechanisms, inventions, and business methods. This Agreement applies regardless of how or to whom a Submission is made.

2. This Agreement takes effect on the date signed by You, and continues until You and COMPANY have signed another agreement.

3. In consideration of Your agreement to be bound to this Agreement's terms, COMPANY is granting You access to its inventor submission application and allowing You to share Your Submissions with COMPANY.

4. This Agreement, Your Submissions, and COMPANY's actions related to Your Submissions: (a) do not create a partnership, joint venture, agency or any other type of relationship not specified in this Agreement; and (b) do not create any financial, equitable or other obligations to You related to any Submissions.

5. You agree that the person signing this Agreement has all necessary authority to enter into, fully perform and be fully bound by this Agreement. You agree that at the time of each Submission: (a) to the best of Your knowledge, You own all rights, title and interest in the Submission.

6. Any actions COMPANY may take related to Your Submission: (a) are not an admission of novelty, uniqueness, priority or originality.

7. Under no circumstance will the COMPANY Parties be obligated to You in any manner based upon Your Submission except under U.S. patent or copyright laws. You waive any other types of claims and actions You may have against the COMPANY Parties related to any Submission.

8. Your Submissions are not made or held in confidence. COMPANY may show them to others (*e.g.*, kids, parents and vendors) who may not have signed a confidentiality agreement.

9. This Agreement: (a) contains the parties' complete agreement regarding the Submissions, (b) supersedes all other agreements, representations and understandings (express or implied). Terms and markings that may accompany Your Submission or related materials do not amend this Agreement, will not be a basis for any waiver of rights under this Agreement, and are not enforceable against COMPANY.

10. This Agreement is governed by and must be construed in accordance with state laws, without regard to its conflict of laws doctrine. The failure or delay by either party to enforce any provision of this Agreement will not constitute a waiver of future enforcement of that provision or any other provision.

COMPANY   INVENTOR
By: _____   By: _____

Name: _____   Name: _____

Date: _____   Date: _____

EXHIBIT D

OPTION AGREEMENT

This Option Agreement is entered into between [INVENTOR COMPANY NAME], hereafter referred to as INVENTOR, a (an) [individual, limited liability company, corporation, etc.], doing business at [Street address, City, State, Zip], and [COMPANY NAME], hereafter referred to as COMPANY, a (an) [individual, limited liability company, corporation, etc.], doing business at [Street address, City, State, Zip].

WHEREAS, INVENTOR has presented to COMPANY, on a confidential basis, the concept known as _____ ™, as defined in the description attached hereto; and WHEREAS, INVENTOR states that it is the inventor of said item and that to the best of its knowledge INVENTOR has all rights pertaining to this concept and has the authority to license said item to COMPANY; and

NOW, THEREFORE, it is agreed as follows:

1. COMPANY agrees to pay a fee of _____ Dollars ($) for an option to develop and manufacture said item.

2. This Option Agreement shall extend from _____ to _____.

3. In consideration for said payment, INVENTOR agrees to refrain from disclosing this concept to any other party, to refrain from licensing the concept to any other party, and further agrees to enter into good faith negotiations with the COMPANY for the licensing of the product to the COMPANY.

4. The COMPANY may disclose this item for the purpose of obtaining manufacturing costs or for the purpose of placing the item with a retailer but all such disclosures shall be made in a confidential basis.

5. Both parties agree to use every effort to speedily conclude the negotiations of a Licensing Agreement on or before _____.

6. Any option fees paid under this Agreement (shall/shall not) be considered royalty advances under any future licensing agreement on said item.

This Agreement is entered into this _____day of _____ 20__.

COMPANY   INVENTOR

By: _____   By: _____

Name: _____   Name: _____

Date: _____   Date: _____

EXHIBIT E

LICENSE AGREEMENT

INVENTOR
And
(Company's name)

**License Agreement
FOR
NAME OF PRODUCT/CONCEPT™**

This agreement made and entered this _____ day of _____, 20___, by and between [INVENTOR NAME], a firm doing business in (STATE), located at (ADDRESS) (hereinafter referred to as "Licensor"), and [COMPANY NAME], a _____ Corporation, having its principal place of business at (ADDRESS), (hereinafter referred to as "Licensee");

WHEREAS, Licensor warrants that it has designed and developed the Licensed Product more particularly defined and identified in Exhibit A; and

WHEREAS, Licensor warrants that it has not granted, and during the term of this Agreement it will not grant any other license or rights in the Territory of the Agreement relating in any way to the Licensed Product; and

WHEREAS, Licensee is desirous of obtaining an exclusive license within the Territory of the Agreement to make, use and/or sell said Licensed Product;

NOW, THEREFORE, in consideration for the undertakings hereinafter set forth, the parties agree as follows:

1. Definition of Terms.

   As used herein, the following terms shall have the following meanings;

   1.1 "Agreement" shall mean this Exclusive Licensed Agreement.

   1.2 "Licensed Product" shall mean the property described and illustrated in Exhibit "A" attached hereto.

   1.3 "Licensee" shall include all of Licensee's subsidiaries, divisions and affiliates now existing or hereinafter formed or acquired.

   1.4 "Net Sales" shall mean the gross sales revenue derived by Licensee from the sale of the Licensed Product and/or products made and sold embodying the Licensed Product less (a) freight (if paid by Licensee) allowances, sales or excise taxes, discounts, returns and uncollectible accounts and recalls or repurchases brought about by or in anticipation of governmental regulations or actions, or (b) _____ percent (_____%) of gross sales, whichever is less. No other cost incurred in the manufacture, sale, distribution or exploitation of the Licensed Product shall be deducted in computing Net Sales.

   1.5 "Related Company" shall mean:

      (i) An organization of which more than thirty percent (30%) of the voting stock is controlled or owned directly or indirectly by Licensee;

(ii) An organization which directly or indirectly owns or controls more than thirty percent (30%) of the voting stock of Licensee; and

(iii) An organization, the majority ownership of which is directly or indirectly common to the majority ownership of Licensee.

1.6 "Territory of the Agreement" shall mean the entire world.

2. License Grant

2.1 Licensor hereby grants to Licensee the exclusive right and license to make, have made, use and/or sell the Licensed Product in the Territory of the Agreement. Licensee shall have the right to sublicense others to make, have made, use and/or sell the Licensed Product in any part of the Territory of the Agreement provided that it has obtained the prior written consent of Licensor which shall not be unreasonably withheld.

2.2 Notwithstanding anything herein contained to the contrary, Licensee shall have the right to contract with another firm or corporation to manufacture the Licensed Product for it, and such subcontractor shall not be liable for any royalty to Licensor. However, in connection with any such subcontract, the Licensee shall require such firm or corporation to acknowledge that its right to manufacture the Licensed Product is derived from this Agreement and to agree to be bound by the provisions of this Agreement (other than the provisions for royalty). Licensee shall in all respects be responsible for such manufacture under all of the terms and conditions of this Agreement with the same effect as if such manufacture were Licensee's own act.

    2.3   Licensee agrees to make a binding commitment for the tooling for the Licensed Product no later than _____, 20___, and to use its best efforts to market the Licensed Product in time for the _____ Season, subject, however, to the provisions of Article 9 hereof relating to occurrences beyond Licensee's control. It is expressly understood and agreed that Licensee's failure to comply with the provisions of this Article 2.3 would be a material breach of its obligations hereunder.

3.   Royalty

    3.1   During the term of this Agreement, Licensee agrees to pay Licensor as royalties, _____ percent (_____%) of Licensee's Net Sales of the Licensed Product and/or products made and sold embodying the Licensed Product and to use its best efforts to maximize such Net Sales. Licensee agrees that in the event that the Licensed Product is sold by Licensee to a Related Company for the purposes of resale, any royalties to be paid in respect to the Licensed Product will be computed on Net Sales of the Related Company when it sells the Licensed Product to its customers in a bona fide, third-party, arms' length transaction.

    3.2   Licensee shall pay Licensor a nonreturnable minimum advance royalty in the sum of _____ Dollars ($_____) with respect to the Licensed Product. Said minimum advance royalty shall be deemed to be advances only against the royalties payable to Licensor with respect to sales of the Licensed Product during calendar years 20_____ and 20_____.

3.3 If during the calendar year 20____, or any subsequent calendar year during the term of the Agreement, royalties payable on domestic Gross Sales (sales within the United States and its Territories) do not equal or exceed a minimum royalty of _____ Dollars ($_____), Licensor shall have the right to give written notice to Licensee of termination of this Agreement. Licensee shall have the right to pay to Licensor the difference between the actual royalties on such domestic Net Sales for such calendar year and the minimum royalty specified in this Article 3.3, within said thirty (30) business days from receipt of said notice. If said difference is paid within said thirty (30) business day period, this Agreement shall continue. If said difference is not paid within said thirty (30) business day period, then this Agreement shall terminate at the end of said period subject, however, to the provisions of Article 13, Rights Subsequent to Termination.

3.4 In the event the Licensed Product in any form is sold FOB a foreign country, then and in that event, Licensee agrees to pay to Licensor a _____ percent (__%) royalty on the Licensee's Net Sales of the Licensed Product and/or products made and sold embodying the Licensed Product.

    3.4-1 If Licensor Product is sold under a nationally recognized license, then a royalty deduction of _____ percent (__%)

shall be in effect for paragraphs 3.1 and 3.4 for those items sold in conjunction with said nationally recognized license.

3.5 Licensor shall have the right to withdraw from the Territory of the Agreement any countries in which Licensee shall not distribute the Licensed Product and/or sublicense others to make, use and/or sell the Licensed Product prior to _____.

3.6 The minimum advance royalty shall entitle Licensee to proceed with the development of the product concept and to one (1) original prototype with renderings and description of item, the receipt of which Licensee hereby acknowledges. It is understood that any further development of the Licensed Product shall be the sole obligation of Licensee. However, such activities shall not result in any ownership rights by Licensee and Licensor shall continue to be the sole and exclusive owner of the Licensed Product. If Licensee shall request Licensor to perform any additional services with respect to the Licensed Product, Licensor shall perform such services and shall charge Licensee its customary rates for such services. Any such additional fees shall not be deemed to be an advance against royalties payable hereunder.

3.7 The Licensee may make modifications or changes to Licensed Product however any changes, modifications or extensions of the product should come under the provisions of this contract.

3.8 In the event that Licensee sublicenses any other person, firm or corporation to make, have made, use and/or sell the Licensed Product in any part of the Territory of the Agreement other than the United States of America and its Territories, Licensee shall pay Licensor fifty percent (50%) of the net income received from such sublicensees. As used herein, the term "net income received from sublicensees" shall mean licensing income, after deduction for registrations, registered user applications, copyrights and copyright applications, patent application, and patent and maintenance taxes, if any.

3.9 In addition to the above, if, during the term of this Agreement, Licensee leases the tooling to others outside of the United States of America and its Territories to enable any such person, firm or corporation to manufacture the Licensed Product, Licensee shall pay to Licensor fifty percent (50%) of the rentals earned by Licensee in connection therewith.

3.10 The payment of Licensor of the royalties, sublicensing income, fees, compensation, and other payments provided for in this Agreement shall be free of any taxes, charges, or remittance fees, whether levied by the Federal, State, or municipal governments in the Territory of the Agreement, or by other authorities, except for such income tax which may be expressly required by the laws of the governments in the Territory of the Agreement to be paid for the account of Licensor. The payment of any such income taxes levied upon or withheld from royalties, sublicensing income, fees, or other payments due to Licensor and the filing of any information or tax returns with respect thereto, shall be the responsibility of Licensee, who shall be liable to Licensor with respect to any amounts, fines, or penalties arising out of or resulting from any failure, delay, or error in discharging the aforesaid obligation.

4. Terms of Payment

   4.1 Said periodic royalty payable under Article 3.1 above shall be payable as follows: Each calendar quarter during the term of this Agreement shall constitute an accounting period (a fractional, initial or terminal period shall be regarded as an accounting period) and Licensee shall, within thirty (30) days after the end of each such accounting period, report in writing to Licensor the total number of units sold and the Net Sales of the Licensed Product during such accounting period. Licensee shall send to Licensor with each such report a check payable to Licensor in payment of the amount due, if any, under this Agreement. All payments made hereunder shall be in the U.S. currency. Time is of the essence with respect to all payments hereunder. Interest at the rate of one and one-half percent (1 ½ %) per month shall accrue on any amount due hereunder from and after the day upon which payment is due until the date of payment. Licensor's receipt of acceptance of any of the statements furnished hereunder or of any royalties paid hereunder (or the cashing of any royalty checks paid hereunder) shall not preclude Licensor from questioning the correctness thereof at any time, and in the event that any inconsistencies or mistakes are discovered in such statements or payments, they shall be immediately rectified and the appropriate payment shall be made by Licensee, with interest as provided for herein.

5. Records

    5.1 Licensee shall keep true and accurate records relating to the sale of Licensed Product under this Agreement to the extent necessary to make the reports and payments provided for herein, and such records shall be open to inspection by Licensor or by a competent third party approved by Licensee, provided, however, that Licensee's approval shall not be unreasonably withheld. Only such records as are necessary in determining the accuracy of the reports and payments rendered by Licensee shall be open for inspection. The inspection provided for herein shall be made during normal business hours, not more than once each calendar year, and only upon at least fifteen (15) days written notice by Licensor to Licensee. All books of account and records shall be kept available for at least five (5) years after the expiration or earlier termination of this Agreement.

    5.2 In the event that Licensor conducts an examination of the royalties payable to it under this Agreement, and as a result of such examination, it is determined that there are unreported royalties payable to Licensor, Licensee shall promptly pay Licensor the unreported royalties, together with the maximum amount of interest thereon allowable by law. Moreover, if the unreported royalties under this Agreement and any other Agreement between Licensor and Licensee aggregate more than Five Thousand Dollars ($5,000), Licensee shall reimburse Licensor for its out-of-pocket costs in conducting such examination as well as the

reasonable attorney fees incurred by Licensor in connection therewith.

5.3 In addition to Licensor's right to examine Licensee's books under Article 5.1 above, Licensee, upon the written request of Licensor, shall furnish to Licensor a statement prepared by the Licensee's chief financial officer certifying as to the accuracy of any royalty statements forwarded by Licensee to Licensor. However, notwithstanding the preceding sentence, in no event will Licensee be required to submit such a statement more than once in any calendar year.

6. Prosecution and Maintenance of Patent Rights and/or Copyrights

6.1 At Licensor's option, it shall apply for design patents and/or copyrights on said Licensed Product in the Territory of the Agreement, if patentable and/or copyrightable, and may proceed with application thereof. In such event, it shall transmit documentary evidence thereof to Licensee. If Licensor fails to file, prosecute or maintain such rights in said Licensed Product within the Territory of the Agreement after requested in writing by Licensee to do so, Licensee shall have the right, but not the obligation, to file, prosecute and maintain said rights in said Territory of the Agreement. However, it is expressly understood and agreed that if no such design patent and/or copyright is obtained, Licensee shall nevertheless be required to make the royalty payments provided for herein. The costs of securing any such design patents and/or copyrights shall be borne by the party making the application.

6.2 In the event that Licensee files for a design patent on the Licensed Product, it shall advise Licensor, in writing, of the date of filing of any such applications for design patents. In addition, within three (3) months after any such filing, Licensee shall advise Licensor, in writing, as to any foreign countries in which corresponding design patent applications are to be filed. Licensor shall have the right to file in other foreign countries at its own expense and in its own name.

7. Infringement of Third Party Patents

7.1 If Licensee is sued for patent infringement by reason of manufacture, sale or use of the Licensed Product, Licensee shall promptly notify Licensor of same. If Licensor does not undertake the defense of said action, Licensee may undertake the defense of any such infringement action; and Licensee shall be allowed to withhold and keep fifty percent (50%) of all future royalties which may become payable to Licensor pursuant to this Agreement until it has recouped fifty percent (50%) of all attorney's fees, court costs, witness fees, travel expenses and damages resulting therefrom and paid by Licensee. Licensor shall have no other obligation to Licensee in connection with any such claim unless it can be shown that any of Licensor's representations and warranties contained in the WHEREAS clauses on page 1 hereof are untrue.

8. Enforcement of Applicable Patents

   8.1   If a third party infringes an applicable patent, which infringement is competitive with Licensee's manufacture, sale or use of the Licensed Product, Licensee agrees to give prompt written notice thereof to Licensor. Licensor may institute an action against said third party, or advise Licensee in writing that it does not intend to do so. If Licensor does not, within sixty (60) days after receipt of said notice, institute an infringement action against said third party or it gives written notice that it does not intend to do so, Licensee shall have the right to institute and prosecute an infringement action against said third party; and Licensor agrees to cooperate with prosecution of said action. The expense and control of the litigation shall be borne and exercised by the party initiating the same, but the other party shall be sent copies of all papers filed therein. The party which has not instituted the action shall have sixty (60) business days after notice of commencement thereof to elect to participate in the litigation. Any net recovery, such as profits, damages, attorney's fees or court costs, shall be divided as follows: fifty percent (50%) to the party starting and paying for the litigation and fifty percent (50%) to the other party, unless otherwise agreed in writing.

9. Impossibility of Performance

   9.1   If acts of God, a public enemy, war, civil commotion, fire, flood, strike, government legislation or control, or other events beyond the control of Licensee prevent its manufacture or sale of the Licensed Product, Licensee's failure on that account to render the reports and make the payments required under Articles 3 and 4 above shall be excused and the minimum royalty called for in Article 3 above shall not be required during such periods of inability to perform.

10. Patent Copyright and Trademark Notices

   10.1 Licensee agrees to mark all products embodying the Licensed Product with the proper patent, copyright and trademark notices. Licensee shall have the right to apply without opposition from Licensor, for trademark registration in Licensee's name and at its expense of any mark by which Licensor describes the Licensed Product. Licensor has not made any search as to the availability of any such trademarks for use in connection with the Licensed Product and makes no representations or warranties as to the successful registration of any such marks. As of the date of this Agreement, there are no patent applications, registered copyrights or trademark registrations or applications with respect to the Licensed Product.

11. Term

   11.1 This Agreement shall continue for the life of the patent or patents that may be granted on the Licensed Product and/or as long as the Licensed Product covered by this Agreement shall continue to be manufactured or sold, whichever is longer, unless sooner terminated under the provisions of this Agreement.

   11.2 Notwithstanding Article 11.1 above, it is understood and agreed that if Licensee ceases the manufacture and/or sale of the Licensed Product for a period of six (6) months, except as provided in Article 9 hereof, Licensor may give written notice to Licensee of its desire to terminate this Agreement for that reason and if Licensee does not, within sixty (60) days from the date of such notice, resume the manufacture and sale of the Licensed Product, this Agreement and the license granted herein shall terminate as of the end of such sixty (60) day period.

11.3 Moreover, notwithstanding anything herein contained to the contrary, if, at any time Licensee decides not to market or ceases to market the Licensed Product, it shall notify Licensor, in writing, of such decision and this Agreement and the license granted herein shall thereupon terminate. However, any such termination shall not affect any of Licensee's liabilities or obligations to Licensor hereunder, including, but not limited to, its obligation: (i) for any unpaid royalties, advances or guarantees; (ii) to use its best efforts to maximize Net Sales: and, (iii) to proceed as required by Article 2.3 hereof.

12. Termination

12.1 If Licensee shall at any time default in making any payment or report hereunder or under any other agreement with Licensor, and shall fail to remedy any such default or breach within thirty (30) days after written notice thereof by Licensor, Licensor may, at its option, terminate this Agreement and revoke the license herein granted to be effective as of the date when written notice to such effect is sent to Licensee.

12.2 Upon the termination of this Agreement, Licensee hereby agrees that the models, tooling, jigs, fixtures and line tooling owned by Licensee (if any) relating to the production of the Licensed Product will be turned over to Licensor, or, at Licensor's option, will be promptly destroyed. Licensor shall have the right to have a representative present to verify any such destruction. In addition, Licensee will confirm in writing that such tooling has been destroyed in accordance with Licensor's written request.

12.3 Except as otherwise provided in Article 12.1 above, after the termination of this Agreement, all rights granted to Licensee hereunder shall forthwith revert to Licensor and Licensor shall be free to license others to make, use and/or sell the Licensed Product and Licensee shall refrain from further use of the Licensed Product or any further reference to it, direct or indirect, in connection with the manufacture, sale or distribution of Licensee's products. Upon the termination of this Agreement, Licensor shall be the sole and exclusive owner of any and all patent applications, patents, and copyrights relating to the Licensed Product.

13. Rights Subsequent to Termination

13.1 Upon the termination of this Agreement under Article 12 above, Licensee shall have the right to sell products under this Agreement then in the process of being manufactured, on hand, or on order, accounting to Licensor for royalties thereon, but shall not have the right to order, manufacture, or sell any other products covered by this Agreement. If, at any time of the termination of this Agreement, there is outstanding, any unexpired sublicense granted hereunder by Licensee, Licensor agrees upon request to continue such sublicense throughout the unexpired portion of its term, provided that Licensee unconditionally assigns to Licensor any and all rights it may have to any income to be derived from such sublicenses.

14. Indemnification

    14.1 Licensee agrees to indemnify and hold harmless Licensor from and against any loss and expense, including attorney's fees, and shall bear all costs in the defense of Licensor arising out of any claims, suits, losses and damages for any alleged defects of said Licensed Product or for any other suits or claims arising out of alleged product liability or any other suits relative and incidental to the manufacture and marketing of said Licensed Product. Licensee agrees to furnish Licensor a certificate of insurance covering said situations.

15. Insurance

    5.1 Licensee agrees that it will obtain, at its own expense, product liability insurance from a recognized insurance company, providing adequate protection (at least in the amount of $1,000,000/$2,000,000) against any claims, suits, loss and damage (including reasonable attorney's fees) arising out of any alleged defects in the Licensed Product. Licensee shall provide Licensor with an insurance certificate evidencing Licensor as an additional named insured. Such certificate shall further provide that it may not be canceled by the carrier without at least thirty (30) days advance notice in writing.

16. Integration, Modification and Waiver

   16.1 All terms and provisions of this Agreement are fully set forth herein or attached as exhibits and no prior understanding or obligation not expressly set forth herein or in such Exhibits shall be binding upon parties and no subsequent modification of this Agreement shall be binding upon the parties unless in writing and duly executed. No waiver by either party of a breach, default or obligation of the other party hereto shall constitute a waiver of any other subsequent breach, default or obligation.

17. Notice

   17.1 For the purpose of any and all communications between the parties with respect to this Agreement including notices, royalty reports and payments, their respective addresses, subject to change upon written notice, shall be:

           INVENTOR
           Street Address
           City, State Zip

   17.2 All royalty reports and payments shall be made to:

           INVENTOR
           Street Address
           City, State Zip

All communications from Licensor shall be addressed to:

>Contact Name
>Title
>Company Name
>Street Address
>City, State Zip

17.3 Any written communication sent by registered or certified mail shall be effective as of the date it is mailed and shall be deemed received as of such date. Written communications sent by any other means shall be effective as of the date it is actually received by the addressee.

18. Assignment

18.1 No Assignment of this Agreement by either party shall be effective unless the parties give their specific prior written consent for such assignment. Notwithstanding the preceding sentence, Licensee may, without such consent, assign this Agreement in connection with the sale or transfer of all or substantially all of its business and Licensor may, without such concern, freely assign the benefits of this Agreement.

19. Venue

19.1 This Agreement shall be governed by the laws of the State of _____, and [STATE] shall be the venue for any action at law or other judicial proceeding for the enforcement of this Agreement. In the event of any action at law or other judicial proceeding for the enforcement of this Agreement, the prevailing party, as determined by the court, shall be entitled to an award of all of its reasonable attorney fees in connection with such action or judicial proceeding.

20. Miscellaneous

   20.1 Section headings contained in this Agreement are for reference purposes only and shall not in any way affect the meaning or interpretation of this Agreement.

   20.2 Upon Licensor's written request, Licensee will provide Licensor with a forecast of its estimated Net Sales for the Licensed Product for the current calendar year and will advise Licensor of any material changes in such forecast. Licensor shall be entitled to make this request only one (1) time in each calendar year during the term of this Agreement.

   20.3 Licensee shall annually send Licensor, without charge, _____ samples of each product embodying the Licensed Product which it is marketing. Licensee further agrees to provide additional samples of the Licensed Product to Licensor, at the lowest prices the Licensee charges its best customers, upon Licensor's written request.

   20.4 Licensee shall send Licensor four (4) copies of each edition of each catalog relating to said Licensed Product.

   20.5 Upon Licensor's written request, at any time, Licensee shall furnish Licensor a statement showing the number of units of the Licensed

Property on hand or in the process of being manufactured.

20.6 Licensee has the right to change the form, shape or function of the said Licensed Product and to produce and sell it under the new form, shape or function, provided, however, that all provisions of this Agreement apply to said changes of Licensed Product.

20.7 Licensor agrees, upon Licensee's request, to execute any and all documents and do all acts necessary to carry out the terms of this Agreement.

20.8 The relationship between the parties established by this Agreement is that of independent contractors. As such, subject to the rights retained or granted to and the obligations undertaken by each party pursuant to this Agreement, each party shall conduct its business at its own initiative, responsibility and expense, and shall have no authority to incur any obligations on behalf of the other party.

IN WITNESS WHEREOF, the parties have executed this Agreement on the date written below.

COMPANY   INVENTOR

By: _____   By: _____

Name: _____   Name: _____

Date: _____   Date: _____

SUMMARY OF CONTENTS

LICENSE AGREEMENT
BETWEEN
INVENTOR AND NAME OF COMPANY

1. CONCEPT: See Exhibit 1

2. ROYALTY PAYMENT ADDRESS:

    INVENTOR
    Street Address
    City, State Zip

3. ADVANCE ROYALTY:

4. OPTION FEE:

5. DOMESTIC SALES ROYALTY:

6. FOB SALES ROYALTY:

7. THIRD PARTY CO. LICENSES:

8. PRODUCT SAMPLES:

9. TOOLING:

10. INITIAL MARKETING:

EXHIBIT 1

EXCLUSIVE LICENSE AGREEMENT
BETWEEN
INVENTOR and Name of Company

DEFINITION OF LICENSED PRODUCT

The written definition of the Licensed Product is as follows:

Name of Prototype

Description of prototype

_____
_____
_____

The terms "Licensed Product" shall mean configurational design(s), functional design(s), process(es), invention(s), patent(s), trademark(s), logo(s), copyright(s) and/or applications in process or prosecution thereof, if any, pertaining to the product as designated herein or such other name(s) as may be used by "Licensee" as defined above. It is expressly understood and agreed that the term "Licensed Product" shall include items which have structural design as defined above and shall be liberally construed.

IN WITNESS WHEREOF, the parties have executed this Exhibit "1" on the date written below.

COMPANY	INVENTOR
By: _____	By: _____
Name: _____	Name: _____
Date: _____	Date: _____

www.ingramcontent.com/pod-product-compliance
Lightning Source LLC
Chambersburg PA
CBHW070300230526
45470CB00002B/658